Francisleile Lima Nascimento

PIBID/CAPES: Practices in Geography, Boa Vista - Roraima

Francisleile Lima Nascimento

PIBID/CAPES: Practices in Geography, Boa Vista - Roraima

High School

ScienciaScripts

Imprint

Any brand names and product names mentioned in this book are subject to trademark, brand or patent protection and are trademarks or registered trademarks of their respective holders. The use of brand names, product names, common names, trade names, product descriptions etc. even without a particular marking in this work is in no way to be construed to mean that such names may be regarded as unrestricted in respect of trademark and brand protection legislation and could thus be used by anyone.

Cover image: www.ingimage.com

This book is a translation from the original published under ISBN 978-613-9-64920-4.

Publisher:
Sciencia Scripts
is a trademark of
Dodo Books Indian Ocean Ltd. and OmniScriptum S.R.L publishing group

120 High Road, East Finchley, London, N2 9ED, United Kingdom
Str. Armeneasca 28/1, office 1, Chisinau MD-2012, Republic of Moldova, Europe
Printed at: see last page
ISBN: 978-620-7-84615-3

FRANCISLEILE LIMA NASCIMENTO

Dedication

I dedicate this research to my father Francisco José do Nascimento and my mother Elielza Lima Nogueira in particular, because through them I am present on this planet. In addition to the upbringing they gave me and for all their dedication. To my siblings Vivian Lima Nascimento and Clayton Lima Nascimento who have been with me in this process of growth and in the fulfilment of yet another stage among others that will come along my life's journey.

There's a phrase I've always liked *from a great inventor, "Our greatest weakness* lies in giving up, the surest way to *-eTcer - try harder -ma -eT' - Thomas Edsn.*

Thank you

I thank GOD, because he gave us life. To my family in general, to my friends at the State University of Roraima - UERR who were always with me, to my friend Lucas Costa Silva, to my friends Olívia Vicente Walker, Elyziane Batista Duarte, Jalsione do Nascimento, Márcia Soares, Thaís Brito Chacon and Laudelina Cruz who translated my abstract into English.

To the professors of the Geography course for all the teaching I did to complete my degree, making me an educator.

In particular, I would like to thank Prof^a. Cândida de Almeida Barbosa Pereira, who is the coordinator of the Degree Course in Geography and the supervisor of this monograph, for all her administrative and pedagogical support during these four years as an academic.

SUMMARY

This research aimed to show how students view the subject of Geography and its importance and the actions developed through the Institutional Teaching Initiation Scholarship Programme - PIBID, by the academic scholarship holders in the subject of Geography at the América Sarmento Ribeiro State School - Boa Vista/RR, where the object of the study was to identify the activities carried out by PIBID in the subject of Geography, with a view to increasing motivation for both students and teachers through innovative methodologies, where students stop being recipients of information and start actively participating in these actions that benefit them and the school as a whole. This programme is a project conceived through the Coordination for the Improvement of Higher Education Personnel - CAPES. Criteria used: The América Sarmento Ribeiro State School was chosen because it participates in PIBID where it has not achieved the performance required at national level by the Federal Government Programme that measures the Basic Education Development Index - IDEB. The target audience for this work was regular high school students in the 1st, 2nd and 3rd grade classes in the afternoon at the América Sarmento Ribeiro State School, located in the West Zone, at Av. Raimundo Rodrigues Coelho, s/n, Senador Hélio Campos neighbourhood in the city of Boa Vista/RR. The data needed to respond to the objectives was obtained from the application of 93 (ninety-three) questionnaires subdivided into 47 (forty-seven) questionnaires focused on knowledge of the subject of Geography and 46 (forty-six) questionnaires focused on the activities developed through PIBID with previously defined closed questions, thus characterising a data survey. The data collection process took place on 2 April 2012, and the results were assessed quantitatively and qualitatively, bringing out both positive and negative points. The set of information was subjected to statistical analysis and testing, with the aim of verifying how the students view the subject of Geography and its importance and the actions developed through PIBID by the academic fellows in the subject of Geography were being worked on, and what methodologies were used so that the result was satisfactory and had an effect on the students' performance and interest in the programme. In general terms, the results obtained from the questionnaires on both the subject of Geography and the PIBID were positive, apart from a few points where more work needs to be done to improve these results. The scholarship students, the students and the head teacher in the subject of Geography through the PIBID brought methodologies that attracted the student's attention for better learning.

Keywords: Teaching. Methodologies. Learning. Geography. Schools. PIBID/CAPES.

SUMMARY

INTRODUCTION

Teaching and learning are part of the same process, and it is the teacher's duty to control this cycle which, through a teaching process, tends to arouse the student's interest. Through learning, the individual assimilates knowledge that throughout their education brings significant results for their external and internal life. As a result, different needs will be shared in order to bring about significant changes in this context, enabling teachers and students to grow personally and intellectually. For the teacher, it is important to enhance their professional CV, and for the student, it is important because of the learning they will acquire; experience; preparation for higher education; as well as starting to build their academic and professional CV.

With the aim of contributing to the improvement of continuing teacher training and deepening relations between the University and the Community, promoting an improvement in the quality of Basic Education, the State University of Roraima - UERR has been running the Institutional Teaching Initiation Scholarship Programme - PIBID - since the second half of 2010, in partnership with the regular high school students of the América Sarmento Ribeiro State School in the city of Boa Vista - Roraima. In this way, PIBID aims to train teachers who are better able to deal with educational problems and seek alternatives to improve teaching and make it more meaningful. It also provides continuing training for teachers, implementing methodological alternatives for teaching, including practical classes (PAJOLA et al., 2009, p.310).

This cycle between the State University of Roraima - UERR, the Institutional Teaching Initiation Scholarship Programme - PIBID and the América Sarmento Ribeiro State School in the city of Boa Vista - Roraima is something that works and is important so that the academic scholarship holders can interact with the regular school students.

Through the PIBID programme, the higher education institution provides an opportunity for academics to put into practice the knowledge they have acquired in the classroom. The school that receives this programme has the opportunity to receive these academics who come with a new baggage and the students have the opportunity to acquire knowledge in a new way, in addition to the school's head teacher having the privilege of having important help in building this learning in the classroom. In addition to forming a team between academics, teachers and students in a school environment, this brings a difference to the improvement of teaching in a school that is in need of this contribution.

Through the actions carried out at the América Sarmento Ribeiro State School in the city of Boa Vista - Roraima by the Institutional Teaching Initiation Scholarship Programme - PIBID, interest was aroused in carrying out this research through the activities carried out by the students in the Geography subject, with the aim of showing the results that the activities developed have brought to the students and the school.

The school is an extremely important place to develop projects, this project is an important methodological practice for the growth of teaching and learning in the process of academic and professional training of future students and intellectual multipliers for the development of a society. It is an environment that needs to develop projects in order to contribute; to achieve goals; to create possibilities that nurture and stimulate the capacity of human beings to satisfy their intellectual needs, as well as benefiting the collective.

The school has a number of factors in hand that directly link this didactic process. This methodological practice is important, as is the content, teaching and learning that prepare this individual. Throughout this work, important themes that are relevant to this research will be addressed. We begin by describing the History of Education in Brazil and its Pedagogical Practices, Education and some of its characteristics, A Passport to Life: Basic Education.

Education in the State of Roraima; National Curriculum Parameters - **PCN's as an instrument of guidelines for teaching, The PCN's of Geography in High School as a primordial element for the construction of new knowledge, The PCN's and their** Legal Bases in High School. Geography as a support in the education of secondary school students, Geography and its Pedagogical Practice; The teaching-learning relationship.

Methodologies and Methods Used for the Development of this Research, Addressing the stages of the research, The target audience chosen, The methods used to collect the data, Treatment of these data obtained.

History and Contextualisation of the América Sarmento State School - Boa Vista/RR, where the work was carried out, Brief History of the city of Boa Vista/RR, History of the Santa Luzia neighbourhood, Location of the América Sarmento State School - Boa Vista/RR, The school and its physical and social aspects.

Coordination for the Improvement of Higher Education Personnel - CAPES, Institutional Teaching Initiation Scholarship Programme - PIBID; The Actions Developed through the Institutional Teaching Initiation Scholarship Programme (PIBID); Analysis and Applicability of the Data Collected, Analysis of the Questionnaires Applied to Regular High School Students in the Subject of Geography, Analysis of the Questionnaires Applied to Regular High School Students through the Institutional Teaching Initiation Scholarship Programme - PIBID and finally the Final Considerations, taking stock of everything that has been explained throughout this work.

CHAPTER 1

HISTORY OF EDUCATION IN BRAZIL AND ITS PEDAGOGICAL PRACTICES

1.1 Education and some of its characteristics

As a result of the growth of cities and capital, education has become more highly valued due to the increase in opportunities in the labour market, thus requiring the qualification of the workforce, with education being the key to this transformation. As a result of this urban and industrial growth, education was split between the elite and the working class, i.e. technical education and university education that met the needs of the technologically advanced. Workers in industry and commerce are distinguished by having a technical education. The interest in education extends to normal schools, the generic name given to courses to prepare teachers (ARANHA, 1996, p.146).

According to Aranha, schools and educators tend to mould the individual into a patriotic citizen, valuing the civic aspect of national consciousness,

> Alongside the expansion of the school network, another objective of educators in the 19th century was to form the national and patriotic conscience of the citizen. Until then, education had had a general and universal character, and now it emphasised the civic aspect, certainly due to the nationalist tendencies of the time (ARANHA, 1996, p.146).

Despite the transformations and facilities in the school network, needs still exist, especially in some developing countries, and even with laws that seek to end or reduce these difficulties, they are still insufficient. Aranha also points out that the growth of the labour market has come about through the expansion of the three educational classes (elementary, secondary and higher), which has led to a diversification of professions, thus improving their expansion,

> The expansion of the three levels (elementary, secondary and higher) of the school network, including the proposal for better integration between them, was due to the expansion of industry and commerce, the diversification of technical professions and bureaucratic staff in business administration and organisation (ARANHA, 1996, p.164).

1.2 A passport to life: basic education

Even though they have access to education, men and women have different access to it, i.e. the majority who benefit are still men, despite the existence of evidence in favour of young people and women having access to knowledge and contributing to society. As a result of improvements in education, the number of students who are no longer illiterate has increased significantly around the world, affecting men and women, children and adults. The balance of the efforts made over the course of the 20th century to increase the possibilities of education is profoundly contrasting: the number of pupils enrolled in primary and secondary schools around the world has risen from around 250 million in 1960 to over a billion today. Despite this, there are still 885 million illiterate people in the world, with illiteracy reaching the following proportion: two out of five women are illiterate and one out of five men is illiterate (DELORS, 2003, p.123-124).

The author emphasises that the teacher is not just an educator, but a friend, a parent, a psychologist, a companion that the student has while they are at school, they are an important part of the process of building their identity in order to face the problems that exist in the school itself and in society in general,

> The teacher's job is not simply to pass on information or knowledge, but to present them in the form of problems to be solved, placing them in a context and putting them in perspective so that the student can make the connection between their solution and other, broader questions (DELORS, 2003, p.157).

In order for teachers to play a "complete" role, they need to work as a team on the activities they carry out. It's a joint effort that brings positive and expected results for the pupils; the school, the teacher, the pupil and the community must and should be linked so that learning is achieved and the pupil and everyone else can acquire new knowledge. Teachers in action with the school and its community can find clues to improving teacher performance and motivation in the relationship they establish with local authorities. When teachers are part of the community in which they teach, their commitment is clearer.

According to Rigal, teachers have a duty to strengthen their autonomy as well as valuing teaching practice, thus improving their working conditions,

> Teachers need to be retrained as professionals and as protagonists. This requalification must include the rational modification of teacher training, the substantive improvement of their working conditions and the elimination of technical control mechanisms, so as to strengthen their autonomy and valorise their practice (RIGAL, 2000, p.190).

CHAPTER 2

EDUCATION IN THE STATE OF RORAIMA

In recent years, the 1st High School Policy has been undergoing a reform that is gradually being consolidated. According to the Portal of the Government of the State of Roraima, 13 April 2009, the first advance was with Basic Education, lasting three years, with the following aims:

2°. Consolidation and deepening of the knowledge acquired in primary school, enabling further study;

3°. Basic preparation for work, exercise of the student's citizenship and continuous learning;

4°. The improvement of the student as a human being, including ethical, critical and social training and the development of intellectual autonomy;

5°. Understanding the scientific and technological foundations of production processes in a praxis relationship, in the teaching of each subject.

The goals that have been constructed so that the teaching-learning process can be applied effectively and that basic education can in fact have a consolidation of knowledge, a preparation for an individual's citizen training and that in this preparation the improvement in this training is critical, with an understanding of the productive processes in teaching in each discipline and especially in geography, is something that is barely perceptible within this school universe, and this is noted through the indices that analyse the progress of a good education, which ends up clashing with the desired goals for excellence in teaching quality.

CHAPTER 3

NATIONAL CURRICULUM PARAMETERS AS AN INSTRUMENT OF GUIDELINES FOR TEACHING

Geography has made great strides, especially in the education of students, which is why there had to be legal backing throughout the country for this progress to continue. Like all subjects, Geography in its educational curriculum has its guidelines, where the National Curriculum Parameters are part of this system, established by the Ministry of Education and Culture - MEC, where teachers and institutions must take part. They are a reference for the teaching-learning process and also for the organisation of this space in the educational system (NASCIMENTO; WALKER; FERREIRA; BEZERRA, 2009, p.131).

The development and consolidation of the National Curriculum Parameters **(PCNs) was based on the** needs of each region with its various problems, with the federal government passing on the responsibility to the public authorities, making them official in each Brazilian municipality.

> The process of drawing up the national curriculum parameters began with the study of curricular proposals from Brazilian states and municipalities, the analysis carried out by the Carlos Chagas Foundation on official curricula and contact with information on the experience of other countries (VESENTINE, 2004, p.17).

"The National **Curriculum Parameters** constitute a quality benchmark for education [...] throughout the country. Its function is to guide and guarantee the coherence of **investments in the education system (VESENTINE, 2004, p.13).**

Today, the reality of geographers as teachers can be said to have been overcome, but not ended, because their curricula are being altered and reformulated for better interaction between teachers and students, revolutionising geography in the context of society. In this constant class struggle, teachers would have the **responsibility and would have to be the main authors for this elaboration of the PCN's,** because they live closely with the reality of their students and not by specialists who don't know the real complexity of the classroom and Brazilian society (NASCIMENTO; WALKER; FERREIRA; BEZERRA, 2009, p.132).

Geography plays an important role in building the knowledge of primary and secondary school students. Like other subjects, geography has the capacity to develop a critical vision in students' lives, where they can evaluate, interpret and understand the world around them. The importance of Geography in teaching is related to the multiple possibilities for broadening concepts,

Geography's contribution to primary and secondary education. Geography, like the other sciences that form part of the primary and secondary curriculum, seeks to develop students' ability to observe, analyse, interpret and think critically about reality with a view to transforming it (VESENTINE, 2004, p.141).

The role of the teacher in the construction of knowledge is extremely important for the development of the society in which he or she is inserted. Teachers are expected to participate in the curriculum parameters, which is not the case at national level, making professional performance difficult. Whatever the methods adopted by the teacher in a classroom, he or she must give the student an overview of what is happening in the world around them. When students receive this knowledge, they evaluate their knowledge and what they have learnt, and within this learning we can mention aspects in the landscape, in space, in the territory, with man himself, among others (NASCIMENTO; WALKER; FERREIRA; BEZERRA, 2009, p.132-133).

"[...] in constructing the concept, the student will confront his or her point of view resulting from common sense and scientific knowledge" (VESENTINE, 2004, p.54).

By interacting with the current news, students encourage their teachers to be aware of everything that is happening in the world on a daily basis. That's why the geography teacher has a great responsibility in the classroom. The study of the subject of Geography over time has been dynamic, for example, research, learning outside the classroom, writing articles, projects, this has led students to have new knowledge and seek more and more within the discipline a specific knowledge making them dedicate themselves more and more (NASCIMENTO; WALKER; FERREIRA; BEZERRA, 2009, p.132-133).

The school's pedagogical political project, as a reference document, should be built collectively, with the participation of all those who are part of the institution. However, this doesn't happen in practice, something that brings great harm to the disciplines, especially Geography (NASCIMENTO; WALKER; FERREIRA; BEZERRA, 2009, p.133).

3.1 The   High School Geography Curriculum Parameters as a key element in the construction of new knowledge

The objectives of the Geography Curriculum Parameters are to articulate pedagogical thinking and geographic thinking, and they hope to strengthen this relationship by favouring reflection on the contradictions that exist in classroom practice. Within the classroom, Geography teaching also has its objectives, such as organising content; the school and teacher must define specific objectives; understanding and interpreting phenomena; mastering graphic languages and recognising references and spatial sets (NASCIMENTO; WALKER; FERREIRA; BEZERRA, 2009, p.133).

Just like teachers, students have their duties within a school institution. The teacher should articulate theory with practice; the teacher plays an important role in everyday school life and is irreplaceable in the teaching-learning process; teaching practice acquires quality when there is the production of knowledge; the teacher's participation in the theoretical-methodological debate is fundamental, enabling them to think about and plan their practice (SECRETARIA DE EDUCAÇÃO BÁSICA, 2008, p.98).

The student, in other words, would be allowed to understand the meaning of citizenship; to exercise their right to interfere in spatial organisation; to have a vision of the complexity of the world; when constructing the concept, the student will confront their points of view resulting from common sense and scientific knowledge. The school institution has the role of developing the political-pedagogical project in the school and in Geography; inserting the project as a curricular component; thinking about pedagogical practices that encourage and stimulate the learning process (NASCIMENTO; WALKER; FERREIRA; BEZERRA, 2009, p.133).

3.1. 1National Curriculum Parameters   : Secondary Education - Legal Basis

For centuries, geography was considered a science of empirical description and served mainly to wage war, in this case by getting to know enemy territories. According to the Ministry of Education and Culture - MEC, the national curriculum parameters have helped to renew this science in primary and secondary education (MINISTÉRIO DA EDUCAÇÃO, 1999, p.70).

From the 20th century onwards, this science began to be renewed and has been undergoing a process of qualification, where criticism has come up against various problems. It has sought to neutralise knowledge in basic and secondary education. Allowing us to analyse the reality of each society in each region with its difficulties (MINISTÉRIO DA EDUCAÇÃO, 1999, p.82).

In the face of these renovations and innovations, the specific role of Geography teaching is to train the student, literating and building in him the competence to guide him in his curricular knowledge structure, starting from various principles, for example, philosophical principles. "Secondary education should guide the training of citizens to learn to live together, to learn to do, and to learn to be [...] transforming **tutored and emphasised individuals into people who can fully exercise their citizenship [...]" (VESENTINE,**

2004, p.62).

Students' education in secondary school must be more complete, they must be led to exercise full citizenship. The revolution in technology and its scientific advances are part of this education. They must be prepared to interpret all these new sources of knowledge through methodologies capable of satisfying local objectives (NASCIMENTO; WALKER; FERREIRA; BEZERRA, 2009, p.134).

CHAPTER 4

GEOGRAPHY AS A SUPPORT IN THE EDUCATION OF SECONDARY SCHOOL STUDENTS

Geography teaching in secondary school concentrates a set of specific features that are important for students to be able to absorb, interpret and understand the spatial transformations that have taken place in different places and times, thus establishing similarities and differences in relation to the changes that have taken place in the municipality, state and country where they live (AVELINO; OLIVEIRA, 2009, p.97).

They must actively participate in the methodological process of building geographical knowledge, reading and interpreting, producing texts and using other resources to record their thinking and geographical knowledge. This doesn't mean that by the end of primary school they will have become geographers, but according to the PCN's, they should be led to examine a theme, analyse and reflect on reality, using different resources and methods from Geography and drawing on the discipline's own way of thinking (PIRES; CONDEIXA; NÓBREGA; DIAS, 2002, p. 176).

In order to realise this process of working with students, it is essential to draw up a plan to establish the objectives and content to be covered, the different discussions on the chosen themes, the ways, possibilities and means of working on them. It is necessary for the teacher to study and reflect collectively with related areas or even individually, in order to choose the object of study that should interest secondary school students and broaden their knowledge of reality.

Geography plays an important role in studying the relationship between man and nature within a historical process, where the formation of society is a consequence of this relationship based on reading and geographical knowledge, especially of the landscape, so that these individuals can transform spaces through the construction of individual or collective projects incorporating a whole set of needs that are part of this space under construction or already built, according to the PCN'S,

> Geography studies the relationship between the historical process that regulates the formation of human societies and the functioning of nature, through the reading of geographical space and landscape. The perceptions, experiences and memories of individuals and social groups are therefore important elements in reading the spatiality of society, in view of the construction of individual and collective projects that transform different spaces at different times, incorporating movement and speed, rhythms and simultaneity, the objective and the subjective, the economic and the social, the cultural and the individual (PIRES; CONDEIXA; NÓBREGA; DIAS, 2002, p.42).

2.1 Geography and its Pedagogical Practice

Citizenship is currently included in the National Basic Curriculum for Education, which foresees a fundamental role for schools. Thus, the school curriculum must act to educate the student as a person, inserted into the society of which he or she is a member, acting in the social, economic and political decisions of his or her country. Citizenship is often misunderstood and the citizen is confused with the consumer who is not satisfied with the service or the product. In reality, it's a process of fighting to have equal conditions when competing in the labour market, and the rights to housing and equality in education and health, among other rights, which is why citizenship is something that is always being sought. The exercise of citizenship, in addition to teaching about rights and duties, must also continue in the struggle so that what has been won is not lost (LIMA; HUERTAS, 2004, p.64).

Most of the time, the geography that is taught in primary and secondary schools, as well as in secondary schools, has almost nothing to do with the geography that is produced in universities at the research level. Most of the time, Geography (or social studies) teachers have not been able to train themselves in a critical process that allows them to become real critics of the textbook. The geography that is taught and learnt no longer motivates them and is certainly far removed from their real needs. Geography has lost what was always special about it - discussing the present reality of peoples, particularly with regard to their spatial context (UMBELINO, 1987, p.270).

The vast majority of schoolteachers know very well that current geography teaching does not satisfy either the student or the teacher. This situation has opened up space for the so-called "textbook industry" to gain ground. Teachers have certainly been the victims of this process. The textbook has become the teachers' "bible" and publishers have not always put books on the market with a minimum of seriousness and scientific veracity. The vast majority contain gross errors, the identification of which could certainly be used to write a book (UMBELINO, 1987, p.241).

The product of academia is emerging through projects, books, articles that are being produced and published, and these are important discussions to add to or change existing conceptions. It's a way of opening up a range of opportunities that allow for the reconstruction of knowledge, especially in Geography, allowing for reflection on what already exists and on new discoveries and trends for methodologies that improve knowledge as a whole and the teaching of Geography (UMBELINO, 1987, p.281).

CHAPTER 5

THE TEACHING-LEARNING RELATIONSHIP

Schools have a duty to play an important role in training citizens to be political individuals, for a substantive democracy that requires them to be protagonists, active and organised. A school that trains citizens has two fundamental objectives: to contribute to the development of a culture of critical discourse on concrete reality on a public level; and to socialise the values and practices of democracy in everyday institutional settings that facilitate active and critical participation and experiences of organisation (BRUNNER, 1985, p.107).

The educational nature of the school is to replace the business view of management as management and the student as a client. In order to facilitate citizen education, democratic spaces and practices must be strengthened, including the participation of different characters in decision-making, and the teacher-student partnership must be strengthened, prioritising the reconstruction of the public sphere and defining an active role for the school institution in its consolidation, a new proposition of links and articulations with the immediate institutional and social context (RIGAL, 2000, p.190).

With the speed of technology and the changes that come with it, there is a huge amount of information that needs to be processed in order for the school to achieve its evolution and also its reconstruction in relation to knowledge, a process that is extremely important for new ways of transmitting knowledge. Because of this evolution, the emphasis that the modern school gave to the processes of instruction and transmission is being questioned, and this analysis needs to be shifted to the processes of knowledge production (how to learn) and knowledge reconstruction (critical re-elaboration) (RIGAL, 2000, p.190).

The school is an important tool where the individual becomes a critical being, rescuing teaching and learning in a collective way,

> This orientation should reinforce the school in its fundamental condition as a critical producer of meaning and help to ensure that pedagogy is not merely a technical-instrumental dimension centred on individual learning. To this end, the importance of teaching-learning processes as instances of collective dialogical production and cultural negotiation must once again be revived (FREIRE, 1985, p.175 apud ROSA, 2000).

The curriculum is the fundamental building block of the school's identity; it is the end product of a set of objects that contribute to individual and social formation,

The curriculum should be considered a cultural product, the nucleus of relations between education, power, social identity and the construction of subjectivity; an institutionalised form for the constitution of subjects, for the production of individual and social identity. As a cultural product, the curriculum is a privileged field in the school for the construction and dispute of hegemony (RIGAL, 2000, p.190).

CHAPTER 6

METHODOLOGY

This research aimed to show how students view the subject of Geography and its importance and the actions developed through the Institutional Teaching Initiation Scholarship Programme - PIBID by the academic scholarship holders in the subject of Geography at the América Sarmento Ribeiro State School - Boa Vista/RR, where the object of the study proposal was to identify the activities carried out by PIBID in the subject of Geography, with a view to greater motivation for both students and teachers, through innovative methodologies, where the student stops being a receiver of information and starts to actively participate in these actions that are for their own benefit and for the school as a whole. Thus providing students with knowledge through pedagogical projects, which is an important methodological practice for growth in their teaching and learning process, leading them to be prepared for higher education and thus contributing to their start in the process of building their academic and professional curriculum.

2.2 Tools used in the research

2.2.1 Exploratory Research

The research was exploratory, making use of a bibliographical survey and questionnaires, in order to become more familiar with the subject in question, seeking to analyse the importance of the Institutional Teaching Initiation Scholarship Programme - PIBID in the student's teaching and learning process.

> This type of research develops studies that give an overview of the phenomenon being studied. As a general rule, an exploratory study is carried out when the chosen topic has been little worked on and it is difficult to formulate and operationalise hypotheses (OLIVEIRA, 2011, p. 54).

6.1.2 Case Study

This is a case study, as it aims to analyse how the Institutional Teaching Initiation Scholarship Programme (PIBID) can influence the lives of students and, above all, how these actions, which start in basic education, are extremely important for those who want to go on to higher education. Pedagogical activities developed from an early age are positive for the construction of knowledge and being able to arrive at higher education with a certain amount of experience that will facilitate their permanence and future intellectual and professional training.

> The case study is a type of research widely used in the biomedical and social sciences. It consists of the in-depth and exhaustive study of one or a few objects, in such a way as to allow a broad and detailed knowledge of them, a task that is practically impossible with other designs already considered (GIL, 2010, p. 37).

6.1.3 Qualitative and Quantitative Methods

The research is based on a qualitative analysis, "A qualitative evaluation is dedicated to perceiving this problem beyond the usual quantitative surveys, which are nevertheless **important" (DEMO, 2008, p. 17). This** type of analysis depends on many factors, such as the nature of the data collected, the size of the sample, the research instruments and the theoretical assumptions that guide the research, so that there is greater knowledge of the object of study and always seeking a greater amount of information in order to have relevant research that achieves the proposed objective.

> The quantitative method means quantifying data obtained through information collected by means of questionnaires, interviews, observations, as well as "the use of statistical resources and techniques ranging from the broadest such as percentage mean, mode, median and standard deviation, to more complex uses such as correlation coefficient, regression analysis" (OLIVEIRA, 2011, p. 27).

In view of this fact, it is essential that, in order to obtain the best performance from any action or project that is being developed, it is necessary to know the object in question and the context in which it is located, so as to be able to propose possible recommendations for the work in general.

6.   2 Survey stages

The initial stage involved a survey of the literature available in public and private libraries in Boa Vista/RR.

In the second stage, a preliminary study of the area where the educational institution is located was carried out on a theoretical basis.

In the third stage, based on theory, a survey was carried out of the historical process of the educational institution, the América Sarmento Ribeiro State School - Boa Vista/RR.

In the fourth stage, based on theory, two questionnaires were drawn up and applied to teachers and students at the educational institution.

In the fifth stage, the data obtained was interpreted and analysed qualitatively and quantitatively, tabulated and presented in graphs expressed as percentages for greater reading comprehension.

In the sixth stage, an analysis was made of the pedagogical actions developed for regular high school students at the América Sarmento Ribeiro State School through the Teaching Incentive Programme -

PIBID in the subject of Geography.

6.   3Target audience

The target audience for this work was students from the Regular High School of the América Sarmento Ribeiro State School, located in the West Zone, at Av. Raimundo Rodrigues Coelho, s/n, Senador Hélio Campos neighbourhood in the city of Boa Vista/RR. It was created by Decree 957 - E/95 of 31 May 1995.

6.   4Data collection techniques

The data needed to respond to the objectives was obtained by applying 93 (ninety-three) questionnaires to the students who are part of the PIBID, where they were subdivided into 47 (forty-seven) questionnaires focused on knowledge of the subject of Geography and 46 (forty-six) questionnaires focused on the activities developed through the Institutional Teaching Initiation Scholarship Programme - PIBID with previously defined closed questions, thus characterising a data survey. The criteria for choosing the questions was in the interest of making a diagnosis of how the students see the subject of Geography and its importance and the actions developed through the Institutional Teaching Initiation Scholarship Programme - PIBID by the academic scholarship holders in the subject of Geography at the América Sarmento State School.

> Among the various definitions of methods, we will consider the most objective to understand that the method is the path that must be taken to achieve the predetermined objectives and the search for possible answers to the initial questions when outlining the problematisation and/or object of study (OLIVEIRA, 2011, p. 19).

The data collection process took place on 2 April 2012, and the results were assessed quantitatively and qualitatively, bringing out both positive and negative points. Through visits to the school, we were able to see how the Institutional Teaching Initiation Scholarship Programme (PIBID) was being developed. The questions were designed according to their specific nature. They are multiple choice questions, with more than one answer marked individually by each student.

6.5 Data processing

The information was analysed in order to see how students view the subject of Geography and its importance, and how the actions developed through the Institutional Teaching Initiation Scholarship Programme (PIBID) by the academic fellows in the subject of Geography were being worked on, and what methodologies were used for the results to be satisfactory so as to have an effect on the students' performance and interest in the programme.

CHAPTER 7

HISTORY AND CONTEXTUALISATION OF THE AMÉRICA SARMENTO RIBEIRO STATE SCHOOL BOA VISTA/RR WHERE THE WORK WAS CARRIED OUT

7.1 A Brief History of the City of Boa Vista/RR

At the time of the creation of the Federal Territory of Rio Branco, the city of Boa Vista was made up of the following neighbourhoods: Porto da Olaria (Francisco Caetano Filho or Beiral), Rói-Couro (São Pedro), Caxangá (the name given to the stream that cuts through the area behind the Military Police Barracks, in Rua Professor Diomedes), Praça da Bandeira and the Centre (FOLHA DE BOA VISTA, 2008, p. 10).

On 9 July 1890, Boa Vista became a city and the new municipality of Boa Vista was created. In the mid-50s, it became the capital of the then Federal Territory of Roraima. The structural changes to the city's urban layout are due to engineer Darcy Aleixo Derenusson, who led a team of the most respected specialists in urban planning, sanitary sewers, rainwater, water supply and electricity with its distribution network (MAGALHÃES, 1986, p. 40).

The city of Boa Vista began as a farm, "Fazenda Boa Vista", founded in the mid-1830s by Inácio Lopes Magalhães. The abundance of water, the natural fields and the buritizais that decorated the local landscape were ideas for cattle ranching. Boa Vista is the capital of the state of Roraima and is located on the right bank of the Rio Branco. It is a planned city with wide streets and avenues all heading towards the centre. Its territorial area is 5,711.9 kilometres^2 (MAGALHÃES, 1986, p. 38).

Pavani and Moura (2006, p.67) report that Derenusson sketched Boa Vista between 1944 and 1946, probably inspired by the cities of Belo Horizonte and Goiânia, the fan shape of Roraima's capital stands out for its radial form.

7.1.1 History   of the Santa Luzia neighbourhood

According to the Roraima state government's public agency, the Roraima Land and Colonisation Institute (ITERAIMA), the entire area was owned by the Prelature of Roraima (now the Diocese of Roraima), and the government was in charge at the time in 1993 (ITERAIMA, 14. Mar. 2010).

The public deed of amicable expropriation, in the form of values accepted by both parties, was drawn up on 30 September 1993, accompanied by the plan and descriptive memorial issued by the Institute for Colonisation and Agrarian Reform (INCRA).

The first dismembered area was equivalent to 2,438,3017 ha, after which another part was dismembered, totalling 439,7043 ha. After all these bureaucratic processes, the process of urbanisation began with the settlements, with those responsible in the administration of the government at the time, Brigadeiro Ottomar de Souza Pinto, who determined the names of the Pintolândia I, II, III and IV neighbourhoods, the latter of which did not come to fruition.

According to the ITERAIMA technician, Mr Assis da Silva, "the neighbourhood with the name of Pintolândia III is today the Santa Luzia neighbourhood", according to Law No. 483 of 9 December 1999, article 31, item XXVII (amends provisions and annexes of Law No. 244 of 6 September 1991).

7.1.2 Location of   América Sarmento Ribeiro State School - Boa Vista/RR

The América Sarmento Ribeiro State School is located in the West Zone of the city of Boa Vista, at av. Raimundo Rodrigues coelho, s/n, Senador Hélio Campos neighbourhood in the city of Boa Vista/RR. It was created by decree 957 - E/95 of 31 May 1995 (fig.01).

**Mapa do Estado de Roraima
Município de Boa Vista - RR**

**Escola   Estadual   América
Sarmento Ribeiro**

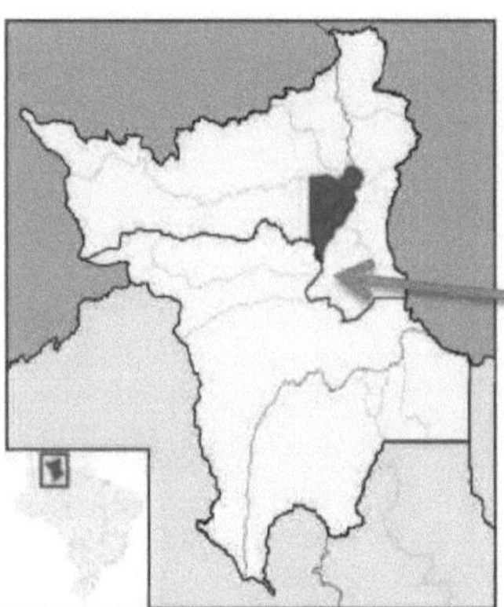

Map of the state of Roraima, highlighting (red) the location of the municipality of Boa Vista-RR; Google Earth satellite image of the América Sarmento Ribeiro State School in the municipality of Boa Vista-RR, 2012.

With the occupation of the neighbourhood, it was renamed Pintolândia III and later Santa Luzia. With the Canaã (Municipal) and América Sarmento Ribeiro (State) schools catering for almost four neighbourhoods, the need arose to build another school that would cater mainly for the Santa Luzia neighbourhood. The idea was spearheaded by Mr César Nogueira Júnior, who, through his perseverance and determination, went door to door to publicise the proposal, mobilised the community and signed a petition, which later culminated in the construction of the school on an unoccupied plot of land used by vandals and squatters who were a nuisance to the local community.

The school under analysis offers public primary and secondary education and serves students from the Senador Hélio Campos neighbourhood, where it is located in adjacent neighbourhoods. The school has the basic infrastructure to cater for its students, with large classrooms without air conditioning, only fans, poorly maintained desks, blackboards and whiteboards, TV and video rooms, computer labs, a sports court, a library, a canteen, a dental room and a reading room. The school has a diverse operational team to meet the demand from its students, including teachers, supervisors, a manager, administrative staff and a general services team.

7.1.3 The   School and its Physical and Social Aspects

The América Sarmento Ribeiro State School currently operates in three shifts with 16 classes in each shift, including primary and secondary education and youth and adult education.

Table 01 below shows the level of education offered by the school, as well as the type of education and the opening hours of each. The data was obtained from the school office.

Table 01 - Level of education offered by the school, as well as the type of education and opening hours for each.

MODALITY	LEVEL OF EDUCATION	TIME
Regular	6º to 8º grade (primary school)	Morning
Regular	1º to 3º grade (high school)	Afternoon
EJA	1º to 3º series of 3º segment	Evening
Special Education	Included pupils in mainstream classrooms	Morning and afternoon

Data provided by the América Sarmento Ribeiro State School - Boa Vista/RR, 2012.

The population that the school serves is basically low-income students with a major deficiency in basic education, according to the level of student learning indicated by the Basic Education Development Index (IDEB). The data used to check the students' level of development and academic achievement was the Provinha Brasil and the School Census.

According to data provided by the school's management, in order to cater for these students, the school has a low demand for teaching resources, such as available and up-to-date books, as well as human resources, since there is a shortage of teachers and to meet this need there are teachers teaching subjects outside their area of training and with an excessive workload (class hours).

Table 02 - Basic Education Development Index - IDEBs observed in 2005, 2007 and Targets for América Sarmento Ribeiro State School.

Elementary School	Observed IDEB		Projected Targets							
	2005	2007	2007	2009	2011	2013	2015	2017	2019	2021
Early Years	-	-	-	-	-	-	-	-	-	-
Final Years	3,3	3,1	3,3	3,5	3,8	4,2	4,5	4,8	5,1	5,3

Data obtained from the website: http://ideb.inep.gov.br/Site/, accessed on 1 June 2010.

These targets are set so that the school can achieve them or even exceed the agreed expectations, but according to the IBED of the América Sarmento Ribeiro State School, it did not achieve what was proposed, with a negative analysis that led the Institutional Teaching Initiation Scholarship Programme - PIBID to become part of this school environment so that it could contribute to improving this index in the quality of teaching.

CHAPTER 8

COORDINATION FOR THE IMPROVEMENT OF HIGHER EDUCATION PERSONNEL - CAPES

The Coordination for the Improvement of Higher Education Personnel (CAPES) plays a fundamental role in the expansion and consolidation of stricto sensu postgraduate programmes (master's and doctoral degrees) in all the states of the Federation. In 2007, it also began to work in the training of basic education teachers, broadening the scope of its actions in the training of qualified personnel in Brazil and abroad. CAPES' activities can be grouped into the following lines of action, each developed by a structured set of programmes (CAPES, 2010):

Evaluation of stricto sensu postgraduate programmes;

- Access to and dissemination of scientific production;

- Investment in the training of high-level resources at home and abroad;

- Promoting international scientific co-operation;

- Induction and promotion of initial and continuing teacher training for basic education in classroom and distance learning formats.

CAPES has been decisive in the successes achieved by the national postgraduate system, both in terms of consolidating the current framework and in building the changes that the advancement of knowledge and the demands of society require. The evaluation system, which is continually being improved, serves as an instrument for the university community in the search for a standard of academic excellence for national master's and doctoral programmes. The results of the evaluation serve as a basis for formulating policies for the postgraduate area, as well as for dimensioning development actions (scholarships, grants, support) (CAPES, 2010).

The National Campaign for the Improvement of Higher Education Personnel (now CAPES) was created on 11 July 1951, by Decree No. 29.741, with the aim of "ensuring the existence of specialised personnel in sufficient quantity and quality to meet the needs of public and private undertakings aimed at developing the country" (CAPES, 2010).

8.1 Institutional Teaching Initiation Scholarship Programme - PIBID

Through these programmes that work to evaluate education in Brazil, whether state or municipal, other programmes have emerged that do not work directly on evaluation, but rather on the process of building knowledge for the purposes of evaluation. One of the programmes working within the school environment is the Institutional Teaching Initiation Scholarship Programme (PIBID), which aims to offer scholarships to higher education undergraduate students to work in public schools. The programme is set up in a school whose IDEB rating is low, i.e. negative, and the school receives the PIBID to improve this index.

The Institutional Teaching Initiation Scholarship Programme - PIBID offers teaching initiation scholarships to students on classroom courses who dedicate themselves to an internship in public schools and who, when they graduate, commit themselves to teaching in the public network. The aim is to anticipate the link between future teachers and public school classrooms. With this initiative, PIBID links higher education (through degree programmes), schools and state and municipal systems (MINISTÉRIO DA EDUCAÇÃO, 2009).

The aim of the programme is to bring together state and municipal education departments and public universities in order to improve teaching in public schools where the IDEB is below the national average of 4.4. Among PIBID's proposals is to encourage teaching careers in the areas of basic education with the greatest shortage of teachers with specific training: science and maths for the fifth to eighth grades of primary school, physics, chemistry, biology and maths for secondary school. In addition to these subjects, others are also chosen depending on the needs and difficulties encountered (MINISTÉRIO DA EDUCAÇÃO, 2009).

Students in higher education have the opportunity to put into practice the theory they have learnt in the classroom, as well as contributing to a better education for students in the public school system. Fellows of the programme work alongside the head teacher of the subject chosen to work in an innovative way, forming a team of educators to transform the classroom into an attractive environment for better learning. The students are responsible for encouraging the teacher to be daring in his or her lessons. They bring playfulness, the use of technology and methodologies that combine theory and practice, making the student seek interest in the subject, identify with it and enjoy learning, giving value to their education.

Federal and State Higher Education Institutions, as well as Federal Institutes of Education, Science and Technology with degree programmes that have been satisfactorily assessed by the National Higher Education Assessment System - SINAES, may submit Proposals for Teaching Initiation Projects. The institutions must have signed a covenant or co-operation agreement with the public basic education networks of the municipalities and states, providing for the participation of PIBID scholarship holders in activities in public schools (MINISTÉRIO DA EDUCAÇÃO, 2009).

The fellows work in the classroom with the teacher, designing activities with the participation of the students. The activities developed are accompanied by the syllabus of the school calendar, but some activities outside the schedule are worked on at the request of the students for reasons of curiosity, lack of knowledge, doubts and other factors that lead to the importance of bringing these difficulties to be developed and that can contribute to the student's knowledge.

These are activities that make use of playfulness, i.e. learning methods that use easy and practical tools to transmit knowledge, often by playing the student is able to grasp the content. In addition to playful activities, field activities are well accepted and well requested by the students, they want to get out of the school space, take advantage of the environment in which they live on a daily basis to learn, practising and experiencing what they have learnt in theory inside the classroom.

The PIBID aims to encourage the training of teachers for basic education, especially for secondary education; to value improving the quality of basic education; to promote the integrated articulation of higher education in the federal system with basic education in the public system, in favour of solid initial teacher training; to raise the quality of academic actions aimed at initial teacher training in degree courses at federal higher education institutions; to encourage the integration of higher education with basic education in primary and secondary education, in order to establish cooperation projects that raise the quality of teaching in public schools; fostering innovative methodological experiences and teaching practices that use information and communication technology resources and are geared towards overcoming problems identified in the teaching-learning process and valuing the public school space as a field of experience for the construction of knowledge in the training of teachers for basic education and providing future teachers with participation in actions, methodological experiences and innovative teaching practices, articulated with the local reality of the school (MINISTÉRIO DA EDUCAÇÃO, 2009).

CHAPTER 9

ACTIONS DEVELOPED BY ACADEMIC FELLOWS IN THE SUBJECT OF GEOGRAPHY THROUGH THE INSTITUTIONAL PROGRAMME OF TEACHING INITIATION SCHOLARSHIPS - PIBID

Before starting the activities through the Institutional Teaching Initiation Scholarship Programme - PIBID, a brief presentation about the programme was given to the students of the América Sarmento Ribeiro State School. The teacher introduced the scholarship students from the Geography course at the State University of Roraima - UERR, to explain why we were there. The scholars were responsible for helping and answering questions together with the teacher with all the students in the classroom.

All the activities were carried out in the subject of Geography. Some activities were carried out using the school's syllabus, where the head teacher sought help in some classes, and others were done for the student's knowledge of the subject of Geography. The activities at the school were carried out as a team - students, teachers and PIBID scholarship students.

One of the activities was on the concepts of Geography. In the textbook, these concepts are explained in a basic way, so we worked on this content in a broader way, first in theory and then in practice. In theory, the main concepts of Geography were explained, which are: place; space; region; territory and landscape.

The classes were dynamic, using technological resources for a better understanding, through images the student can better understand what is being taught. After the theoretical lessons, the students put what they had learnt into practice. Through collage using magazines, the students made posters identifying each concept studied. They represented each concept on a specific poster, using pictures to illustrate all the concepts, as well as putting what they had learnt into their own words.

Another activity was to explain what a Permanent Preservation Area (PPA) is, a subject that is discussed on a daily basis and which has recently changed. It's a topical subject and it's extremely important for students to be aware of it. In addition, in the area where the school is located, there are several APP areas, and this further reinforces the importance of showing and explaining this subject.

Firstly, we worked on the theory, explaining what an APP is, using innovative tools and showing images to better understand the content. After the theory, there was practice. A day was set aside for a field trip to show an APP area located in the neighbourhood itself.

Many students live in the same area, where some are on the edge of an APP area. It was a lesson that brought a better understanding to the students, they loved it, reporting that experiencing what they had learnt in practice was much better. After the field lesson, the students put what they had learnt into practice. They built two models talking about APP, one model was of what an APP area would look like according to environmental laws and the other was a model of what an illegal APP area would look like, in other words, how it is being used, totally irregularly.

Every school has a special date when a cultural fair takes place. During the time that PIBID was at the school, this fair was held, and the scholarship students in the Geography subject through PIBID took part in this fair. All the activities developed through the programme together with the students and the teacher were exhibited at the fair. The students liked the work and were able to share it with all the students at the school.

Activities like these that have been developed at the school are of great importance in arousing student interest, whether in the subject of Geography or in other subjects. Students don't ask for big lessons, they want to learn in an easy way, in a way that fits in with their learning. Theory has to go hand in hand with practice, and the main thing is to get students into that practice. Listening to students' suggestions is very important, it's a way of encouraging them to seek knowledge in a didactic way. It's important for both the student and the teacher.

CHAPTER 10

RESULTS OF DATA APPLICABILITY

10.1 Data analyses

The data collection process took place in April 2012, and the results were assessed quantitatively and qualitatively, bringing out both positive and negative points.

The América Sarmento Ribeiro State School caters for primary and secondary school students. The students chosen were those who attend regular secondary school in the 1st, 2nd and 3rd year classes in the afternoon and who take part in the activities developed by PIBID.

10.1.1 Questionnaires Applied to Regular High School Students in the Geography Subject

A total of 47 (forty-seven) questionnaires were administered to the students, with both boys and girls aged between 17 and 18 taking part. The questionnaires were administered in the afternoon. In the 1st, 2nd and 3rd year of regular secondary school, the questionnaires were administered to one class in the 1st year, one class in the 2nd year and three classes in the 3rd year.

The questionnaire was designed to gain a better understanding of the teaching of Geography, how the students view the subject, how the Geography teacher relates to the students in the classroom, whether the student is able to understand the content, what suggestions the students would like the teacher to put into practice in the subject of Geography and how important Geography is in the professional and intellectual life of these students.

The questions are multi-choice questions, with more than one answer marked individually by each student.

Given these changes in the way Geography is taught, research such as this is of great importance so that education professionals, especially Geography teachers, can analyse what kind of knowledge secondary school students are receiving, and whether these changes are positive or negative for the teaching-learning process of these students during this cycle of secondary school education. It's important to know how Geography is present in these individuals. What can be done to improve and encourage students to enjoy geography?

The survey of 93 interviewees in the 1st, 2nd and 3rd year of secondary school in the afternoon shows that 47% of the students interviewed at América Sarmento Ribeiro State School are female and 53% are male. Despite the fact that, according to the IBGE (2007, 20 August), women make up more than half of the Brazilian population, according to the Synthesis of Social Indicators - 2007, the Brazilian population in 2006 was 187.2 million inhabitants, 96 million of whom were women. Based on IBGE data, the number of women attending an educational institution is still lower than the opposite sex (graph 01).

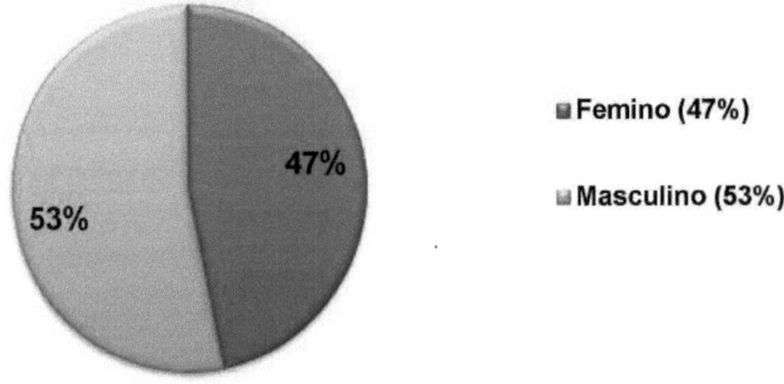

Graph 01
Sex

The school institution has the role of developing the Political-Pedagogical Project in the school and in Geography; inserting the project as a curricular component; thinking about pedagogical practices that encourage and stimulate the learning process.

The students were then asked how they would define the subject of Geography in general, analysing how important Geography is for their lives. Of these, 23% chose to say that it is important for their own knowledge; 22% said that it is a very broad subject of knowledge; 33% said that it is present in everything that happens in the world; 21% consider it a subject that is constantly changing and 1% said that it is just an easy and decorative subject, having no importance for academic and professional growth, which shows that the students have the concept that Geography is part of our daily lives and the constant dynamics of our planet (graph 02).

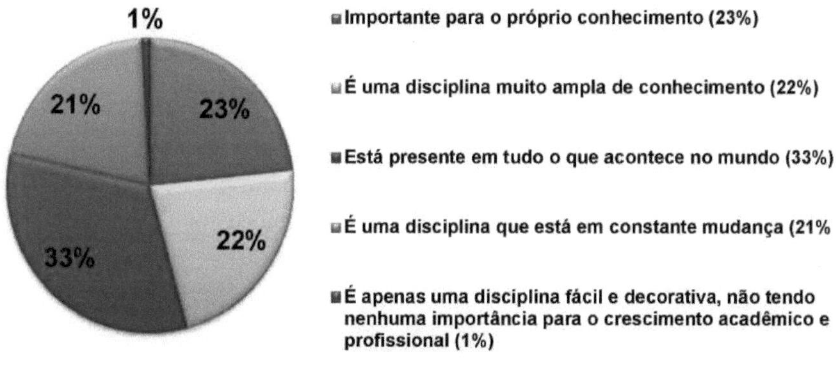

Graph 02
How would you define the subject of GEOGRAPHY today?

The role of the teacher goes far beyond just teaching, he is part of the student's life, he is responsible for the knowledge he transmits and the relationship he establishes with this individual, the respect between teacher and student, the friendship that is built, the trust that the teacher needs to pass on to his student, there are several factors that contribute to a bond of commitment. Concerned about this, we asked about the relationship between teacher and student in the classroom. 51 per cent of those interviewed said that the relationship between teacher and student was excellent; 38 per cent said they had a good relationship; 9 per cent said they had a fair relationship and 2 per cent had a poor relationship with the teacher (graph 03).

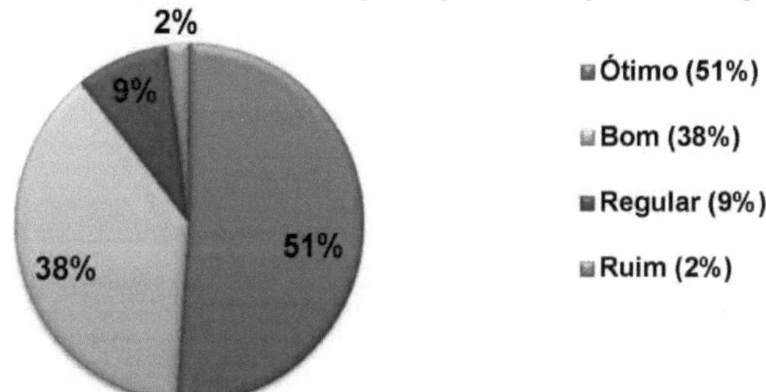

Graph 03
How is the relationship between the GEOGRAPHY teacher and the students?

Whatever the methods adopted by the teacher in a classroom, he or she must lead the student to have an overview of the things that happen in the world around them. When students receive this knowledge, they evaluate their knowledge and what they have learnt, and within this learning we can mention the aspects of the landscape, space, territory and man himself (NASCIMENTO; WALKER; FERREIRA; BEZERRA, 2009, p.132).Based on this, it is important to know if the student is understanding the content that the teacher teaches in the classroom. When asking the students, 89% said that they understand what the teacher teaches in the classroom and 11% said that they do not understand the content that the teacher passes on in the classroom. This shows that within the universe surveyed, with 47 interviewees focused on this issue, we can conclude that the percentage of 11% is very high and is of extreme relevance and concern when it comes to learning (graph 04).

Graph 04
Can you understand the content presented by your GEOGRAPHY teacher?

In an attempt to find out about the teacher's conduct in various aspects, we looked at the teacher's negative points in the classroom. Graph 05 shows that 7% of those interviewed said that the teacher doesn't master the Geography content; 7% said that the teacher doesn't know how to explain the Geography content; 15% said that the teacher talks about other subjects in the classroom; 5% said that the teacher only gives tests in the Geography subject; 6% said that the teacher is only late for Geography lessons; 6% said that the teacher doesn't teach the 60-minute lesson in Geography; 27% said that the teacher only uses the textbook in Geography; 21% said that the teacher doesn't use didactic lessons, such as dynamics in the classroom, and doesn't use games to bring other learning results to their lessons; 6% said that the teacher doesn't have a degree in Geography.

In general terms, we can see that two results were considered to be somewhat negative, the first being the lack of more practical lessons with new learning methods and the sole use of textbooks in the classroom. Unfortunately, as the years go by, teachers lose the essence of their role in the classroom and this leads to monotonous, repetitive, tiring lessons that don't interest students in the subject, so the use of methods that bring satisfactory results in the student's teaching and learning is important, the teacher has to stop being absorbed by the textbook, use other references and technologies that help in the contact of new knowledge and subjects of daily life, current affairs that are part of the student's life and that should be used in the classroom for a better understanding, especially in the subject of Geography, which is constantly changing (graph.05).

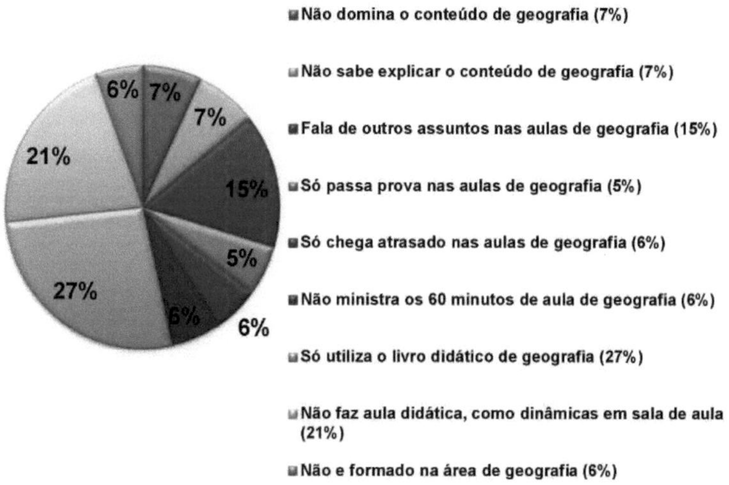

■ Não domina o conteúdo de geografia (7%)

■ Não sabe explicar o conteúdo de geografia (7%)

■ Fala de outros assuntos nas aulas de geografia (15%)

■ Só passa prova nas aulas de geografia (5%)

■ Só chega atrasado nas aulas de geografia (6%)

■ Não ministra os 60 minutos de aula de geografia (6%)

■ Só utiliza o livro didático de geografia (27%)

■ Não faz aula didática, como dinâmicas em sala de aula (21%)

■ Não e formado na área de geografia (6%)

Graph 05

In relation to the GEOGRAPHY teacher in the classroom, what are the NEGATIVE points?

In an attempt to find out about the teacher's conduct in various aspects, 12% pointed out that the teacher has a degree in Geography; 18% said that the teacher knows how to explain the Geography content; 7% said that the teacher uses other tools besides the Geography textbook; 8% considered that the teacher gave didactic lessons, such as dynamic lessons in the classroom; 8% said that the teacher gives field lessons in Geography; 16% said that the teacher gives seminars to the students in Geography lessons; 10% said that the teacher gives 60-minute lessons in Geography; 9% said that the teacher is punctual in Geography lessons and 12% said that the teacher masters the Geography content.

In this process, a percentage will define how some aspects related to classroom teaching are going, and how a teacher, especially in the subject of geography, behaves and exercises their role as an educator, being a reference for their students. It is extremely important for educators to be able to work and develop the assessment process well in the construction of knowledge. This assessment is useful and must be constant in the growth of teaching and learning, enabling students to achieve positive results that will lead them to understand and interpret what is being transmitted.

When this evaluation process is not well constructed, through a survey such as this one, the student's lack of understanding of what is being asked of him is noticeable, because with the result acquired we can analyse that the confrontation in the answers presented, and this conflict of knowing what is being answered and affirming, is worrying in the sense of the research analysis itself. So this leads us to think about how the teacher is being responsible or not in this evaluation process that is important in this construction of knowledge.

Overall, we can see that teachers need to innovate in their lessons, especially in Geography. This percentage needs to increase. This survey will be important for teachers to realise that changes need to happen and that students want them to (graph 06).

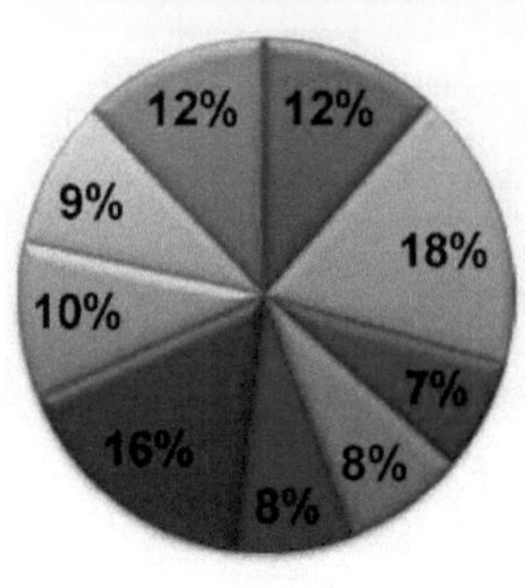

Graph 06

What are the POSITIVE points about the GEOGRAPHY teacher in the classroom?

The points raised, both negative and positive, were important because they allowed the students to say what is lacking in Geography classes and what the teacher needs to do to arouse student interest in this subject. Some suggestions were raised so that the students could point out what they were interested in developing in the classroom: 19% of those interviewed said they would like to see field lessons in Geography; 12% said they would like to see classroom dynamics in Geography; 13% said they would like the teacher to use maps in Geography; 17% think that the use of technology in geography is important; 13% said that they would like to work with research in geography; 12% said that they would like to work with theory and practice in geography and 14% said that they would like the teacher to use tools other than the geography textbook (graph 07).07.

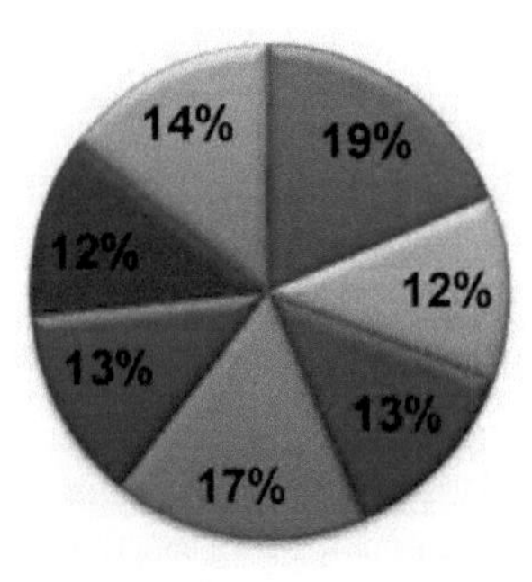

Graph 07

What suggestions would you make to the GEOGRAPHY teacher for developing work in the subject?

Of the students interviewed, 8 per cent said that Geography was important for their intellectual training; 17 per cent said that Geography was important for their professional training; 30 per cent said that Geography was important for their intellectual and professional training; 2 per cent said that Geography was not important for their intellectual training; 2 per cent said that Geography was not important for their professional training; 2 per cent said that Geography was not important for their intellectual and professional training; 3 per cent said that there was no need to study Geography and 36 per cent thought that studying Geography was essential.

Although the result was a sample, it was satisfactory. Geography is valued, and the attention it receives can be increasingly reinforced. The role of the Geography teacher in the classroom as an educator is up to him to show the student that the study of Geography is important for the intellectual and professional formation of every human being (graph 08).

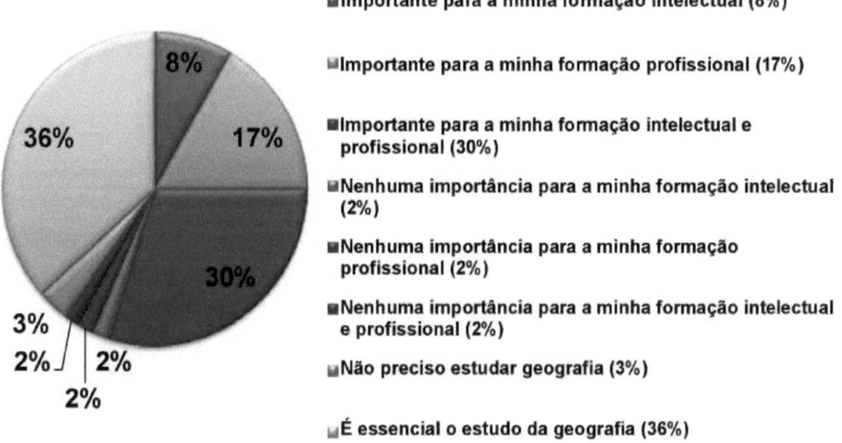

Graph 08
How important is GEOGRAPHY in your intellectual and professional development?

10.1.2 Questionnaires applied to regular secondary school students through the Institutional Teaching Initiation Scholarship Programme - PIBID

A total of 46 (forty-six) questionnaires were administered to students in the afternoon. In the 1st, 2nd and 3rd year of regular secondary school.

This study presents an analysis carried out to diagnose student knowledge of the subject of Geography and the activities developed through the Teaching Incentive Programme - PIBID at the América Sarmento Ribeiro State School, a survey carried out through questionnaires applied to students in order to find out how the subject of Geography is present in the intellectual and professional lives of these students and to find out the results that PIBID in the subject of Geography has brought to them.

The data collection process took place in April 2012, and the results were assessed quantitatively and qualitatively, bringing out both positive and negative points.

The questions are multi-choice, with more than one answer marked individually by each student.

Through these programmes that work to evaluate education in Brazil, whether state or municipal, other programmes have emerged that do not work directly on evaluation, but rather on the process of building knowledge for the purposes of evaluation. One of the programmes working within the school space is PIBID, which aims to offer scholarships to higher education undergraduate students to work in public education networks. The programme is set up in a school whose IDEB rating is low, i.e. negative, and the school receives PIBID to improve this index.

One of the objectives of PIBID is to encourage the integration of higher education with basic education in primary and secondary schools, in order to establish cooperation projects that improve the quality of teaching in public schools. The students interviewed were asked about their knowledge of the programme. 21% said yes, they were aware of the programme in their school, while 79% said no, they were unaware of the programme's existence. Unfortunately, the students are unaware of the importance of PIBID in their school. Despite the presentation made and the activities that were carried out, the number of students who did not associate the programme with the work carried out was very high and this is worrying in terms of how PIBID is actually involving the student in its activities (graph 09).

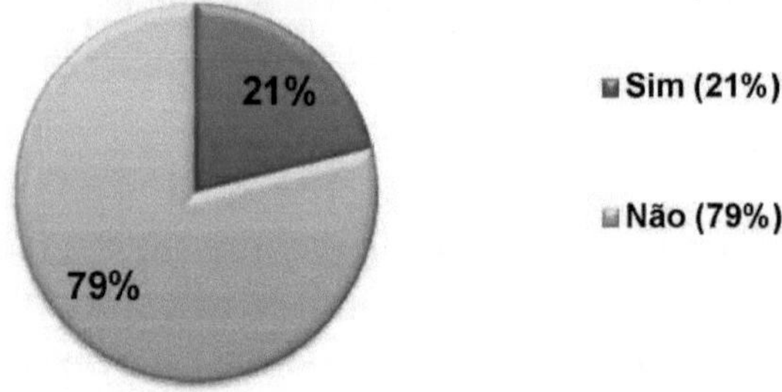

Graph 09
Are you aware of the Institutional Teaching Initiation Scholarship Programme (PIBID) at your school?

The fellows work in the classroom with the teacher, designing activities with the participation of the students. The activities carried out are monitored through the syllabus of the school calendar, but some activities outside the schedule are worked on at the request of the students for reasons of curiosity, lack of knowledge, doubts and other factors that lead to the importance of bringing these difficulties to be developed and that can contribute to the student's knowledge. The students interviewed were asked if they liked the work carried out by the scholarship students in the subject of Geography through PIBID, 53% said yes, they liked the activities carried out through the programme, 47% said no, they did not like the actions carried out by the scholarship students in the subject of Geography at school. Unfortunately, this is a worrying result, as the fellows, together with the head teacher, did not achieve the programme's objectives, leaving the students dissatisfied with the performance of the activities they worked on (graph 10).

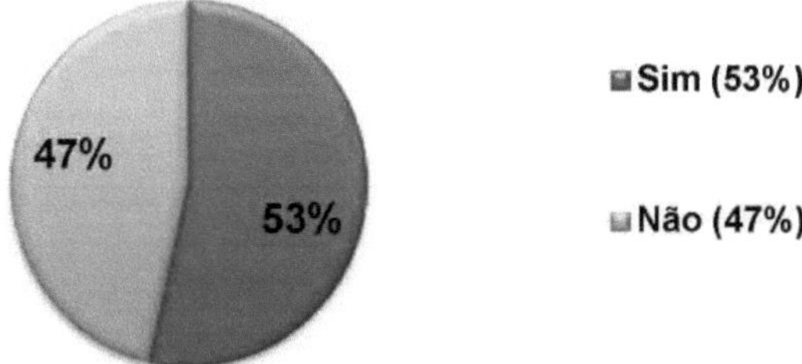

Graph 10
Did you like the work carried out by the geography students through the Institutional Teaching Initiation Scholarship Programme - PIBID at your school?

In addition to the role of educator that the scholarship holder is developing at the school together with the students and the head teacher, his relationship with them is also of great importance for better results in the activities. When the students interviewed were asked about the relationship between the students and the academic fellows in the Geography course, 30% said that the relationship was excellent, 36% said that the relationship was good, 27% thought that the relationship was regular and 7% said that the relationship between the students and the academic fellows was bad. Overall, the academic fellows have a good relationship with the students, which shows that there is a rapport, respect and trust in the work they do at school (graph 11).

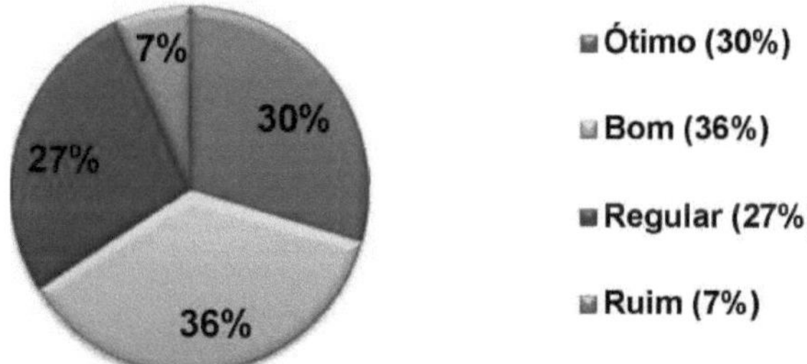

Graph 11

In your opinion, was the relationship between the students and the academic fellows of the GEOGRAPHY course good?

With regard to student learning, the students interviewed were asked about their opinion on the progress made in their learning after the activities developed by the academic fellows in the subject of Geography through PIBID. 49% said that yes, they noticed a difference in their learning after the programme was set up in the school and its actions were initiated, 51% said that no, their perception of the actions did not bring any improvement in their knowledge, they did not notice any change in their teaching-learning process.

A negative result, because the majority did not notice or did not have any improvement in their learning through the activities developed by the scholarship students through PIBID in the subject of Geography, the academics together with the head teacher did not reach a goal in which more students could perceive such changes, this shows that the scholarship students and the teacher need to identify the flaws and improve them so that the results are aimed at on a larger scale, and that the students, the greatest and important beneficiaries of this programme, will have better learning, making everyone's efforts worthwhile for a better education and educational growth (graph 12).

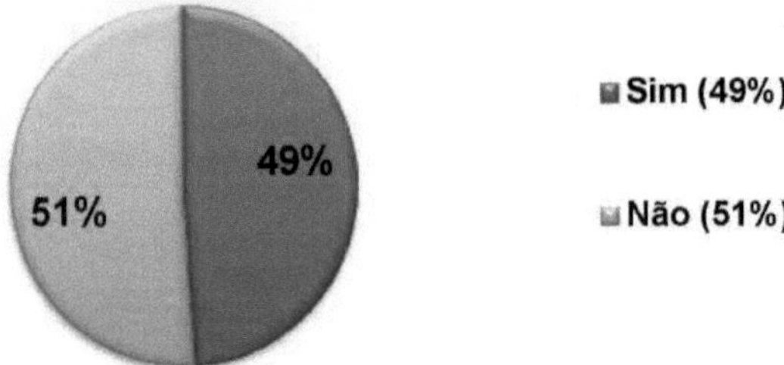

Graph 12

In your opinion, has there been any progress in teaching and learning after the activities developed by the academic fellows in the subject of GEOGRAPHY through the Institutional Teaching Initiation Scholarship Programme - PIBID?

The students are responsible for introducing the teacher to play, the use of technology and methodologies that combine theory and practice, making the student seek interest in the subject, identification, the pleasure of learning, giving value to their education. The students interviewed were asked to identify the activities developed by the academic fellows through PIBID in the subject of Geography that were most important. 26% said they had field lessons in the subject of Geography, 11% said they had classroom dynamics in the subject of Geography, 15% said they had research in the subject of Geography, 11% said they had theory and practice in Geography lessons, 9% said they built models in Geography lessons, 28% said that other activities not mentioned were developed in the subject of Geography. The practical and playful activities were well explored, which shows the importance of continuing with these actions and improving them more and more so that students become interested in studying and in the subject of Geography (graph 13).

Graph 13

Could you identify the most important activities carried out by the PIBID GEOGRAPHY scholarship holders?

Through these activities developed by PIBID in the subject of Geography, the students interviewed were asked if they would like these actions to remain in their school for their benefit. 76% said yes, they would very much like these activities to continue, while 24% said no, they have no interest in the actions developed by the academic fellows in the subject of Geography through the programme. This result is positive, it shows that the programme is being recognised and that it is playing an important and significant role in the teaching-learning process of these students who benefit from the programme (graph 14).

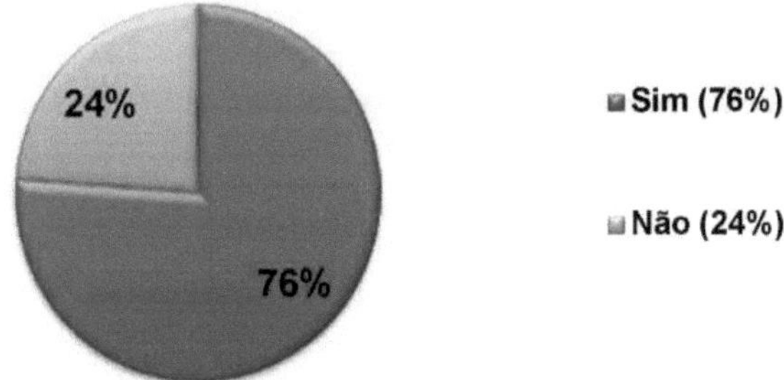

Graph 14

Would you like the activities of the Institutional Teaching Initiation Scholarship Programme (PIBID) developed in the subject of GEOGRAPHY to remain at the school?

Through these actions developed by PIBID in the subject of Geography, we asked the students interviewed if they became interested in the subject after the activities started through the programme in their school. 39% said yes, there was a greater interest in the subject of Geography and 61% said no, the programme did not change their opinion or interest in the subject. Unfortunately, this result is sad, because the programme is a way of awakening the student's interest in the subject of Geography. Although the activities are didactic and modify the usual day-to-day lessons, they often don't change the student's taste or identification, Each student has an identity with each subject in particular, those who are part of the world of Geography try to make the student sympathise with it, but this is not always possible, however achieving 39% student interest in the subject is very important, as it is a way for us educated in particular the Geography teacher to dedicate ourselves even more to it (graph.15).

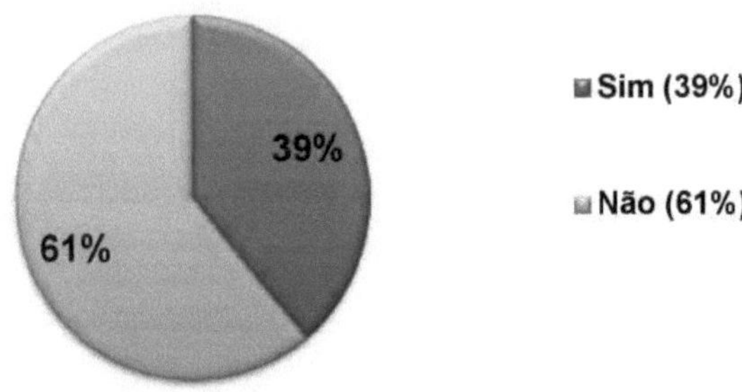

Graph 15

Did you become interested in the subject of geography after the PIBID/GEOGRAFIA programme started at your school?

Looking for suggestions of activities that PIBID scholarship students in the subject of Geography could be acquiring, the students listed that they would like to see more developed, actions such as field classes; that the teacher doesn't just use the textbook; more dynamic classes; construction of models; creative seminars; research work; theory and practice in Geography classes. These are important suggestions that the students would like the teacher to pay attention to and think about in an important way for the construction of knowledge and growth in the student's teaching and learning. It's a way of exploring the student's knowledge, stimulating their curiosity about knowledge and being able to have meaningful and important experiences for their future training. These are actions that are extremely important for those who want to go on to higher education. Certain activities developed from an early age are positive for the construction of knowledge and being able to arrive at higher education with a certain amount of experience that will facilitate their permanence and future intellectual and professional training.

CHAPTER 11

FINAL CONSIDERATIONS

The research shows that through the teaching-learning process of the work carried out in this school environment, it is possible to learn about the reality of teaching in the subject of Geography and how PIBID was developed in the school. In view of the results obtained, it was observed that new teaching proposals for the institution, new projects and methodologies for the real adaptation and assimilation of content is something necessary to bring positive results in learning, with students expressing this concern.

All the activities were carried out in the subject of Geography. Some were carried out through the school's syllabus, where the head teacher sought help in some lessons, and others were done for the student's own knowledge. The activities at the school were carried out as a team - students, teachers and PIBID scholarship students.

Pedagogical practices such as these developed at the school are of great importance in arousing student interest, whether in Geography or other subjects. Students want to learn in a relaxed way, in a way that fits in with their learning. Theory has to go hand in hand with practice, and the main thing is to get students into that practice. Listening to students' opinions and suggestions is very important, it's a way of encouraging them to seek knowledge in a didactic way. It's important for both the student and the teacher. Everyone wins with this: the school, the teacher and the student.

In this process, a percentage will define how some aspects related to teaching practice in the classroom are going, and how a teacher, especially in the subject of geography, behaves and fulfils their role as an educator, being a reference for their students. It is extremely important for educators to be able to work and develop the assessment process well in the construction of knowledge. This assessment is useful and must be constant in the growth of teaching and learning, so that the student can achieve positive results that will lead them to understand and interpret what is being transmitted.

When this evaluation process is not well constructed, through a survey such as this one, the student's lack of understanding of what is being asked of him is noticeable, because with the result acquired we can analyse that the confrontation in the answers presented, and this conflict of knowing what is being answered and affirming, is worrying in the sense of the research analysis itself. So this leads us to think about how the teacher is being responsible or not in this evaluation process that is important in this construction of knowledge. As a suggestion, we could conduct a study into how this assessment process is being applied to the construction of knowledge.

An analysis of the results obtained both from the questionnaires on the subject of Geography and from the Institutional Teaching Initiation Scholarship Programme (PIBID) brought positive results, apart from a few points where more work needs to be done to improve these results. The scholarship students, the students and the head teacher in the subject of Geography through PIBID brought methodologies that attracted the student's attention for better learning.

But this doesn't mean that some points of the research didn't give rise to concern. On the contrary, these points, which in a way are "negative", are valid for improving the performance of teachers and scholarship students so that in other actions the commitment is greater and the objectives are more successfully achieved.

REFERENCES

ARANHA, Maria Lucia de Arruda. **History of education**. 2. ed. rev. and current. São Paulo: moderna, 1996.

AVELINO, Patricia Helena Mirandola; OLIVEIRA, Arlinda Montalvão de. **The teaching of physical geography content in youth and adult education - EJA:** experience reports of a pedagogical practice applied at Bom Jesus State School - Três Lagoas-MS. In: 10th National Meeting of Teaching Practice in Geography (ENPEG), Porto Alegre, UFMS, 2009.

BRAZIL. Ministry of Education. Secretariat of Media and Technological Education. **National Curriculum Parameters:** Secondary Education - legal bases. Brasília, 1999.

BRUNER, Jerome. **Education, the door to culture**. Madrid. Visor. 1997. In: Education in the 21st century: the challenges of the immediate future, ed. Porto Alegre: Artes Médicas Sul, 2000.

BRUNER, Jerome. **Education and the future of democracy**. Working document n73. Santiago de Chile. Flacso. 1985. In: Education in the 21st Century: The Challenges of the Immediate Future, ed. F. Imbernón; trad. Ernani Rosa. 2.ed. Porto Alegre: Artes Médicas Sul, 2000.

COORDINATION FOR THE IMPROVEMENT OF HIGHER EDUCATION PERSONNEL (CAPES). Available at: <http://www.capes.gov.br/index.php>. Accessed on: 07 March 2010.

DELORS, Jacques. **Education: a treasure to be discovered:** report to Unesco by the International Commission on Education for the 21st Century. 8.ed. São Paulo: Cortez; Brasília, DF, MEC: UNESCO, 2003.

DEMO, Pedro (1941). **Qualitative evaluation**. 9. Ed. - Campinas, SP: Autores Associados, 2008, (collection polêmicas do nosso tempo; 25).

FREIRE, Paulo apud ROSA, 2000. **Interview with Paulo Freire**. Buenos Aires. CEDAL, 1985. In: a educação no século XXI: os desafios do futuro imediato, ed. F. Imbernón; trad. Ernani Rosa. 2.ed. Porto Alegre: Artes Médicas Sul, 2000.

GIL, Antonio Carlos, 1946. **How to prepare research projects**. - 5. ed. - São Paulo: Atlas, 2010.

BRAZILIAN INSTITUTE OF GEOGRAPHY AND STATISTICS. IBGE. Available at: <http://www.ibge.gov.br>. >. Accessed on: 20 August 2010.

RORAIMA LAND AND COLONISATION INSTITUTE - ITERAIMA. Ville Avenue Roy, 1500 Canarinho. Boa Vista, Roraima, Brazil. Institutes and Foundations. Available at: <http://www.iteraima.rr.gov.br>. >. Accessed on: 14 March 2010.

LIMA, Alessandra de; HUERTAS, Maria Del Carmen Matilde. **Geography and citizenship in secondary education.** In: multiple geographies: teaching - research - reflection, eds. ASANI, A. Y; ANTONULO, I. T; TSUKAMOTO, R. Y. Edições Humanidades, Londrina, 2004.

MAGALHÃES, Dorval. **Roraima Historical Information**, Boa Vista, 1986.

MINISTRY OF EDUCATION. **Institutional Teaching Initiation Scholarship Programme - PIBID,** 2009. Available at: <http://portal.mec.gov.br/index.php?Itemid=467&id=233&option=com content&view=article>. Accessed on: 07 March 2011.

NASCIMENTO, Francisleile Lima; WALKER, Olívia Vicente; FERREIRA, Sebastião Lopes; BEZERRA, Josinaldo Barboza. **Curriculum Parameters for Geography in Secondary Education**. Norte Cientifico, v.4, n.1, Dec. 2009, Boa Vista, RR, IFRR, 2009.

OLIVEIRA, Maria Marly de. **How to write projects, reports, monographs, dissertations and theses.** - 5. ed. [rev.] - Rio de Janeiro: Elsevier, 2011.

PAJOLA, F.; BARBOSA, R.U.; SOUSA, D.N.; GONÇALVES JUNIOR, O.D.; CARVALHO, F.A.; OLIVEIRA, H.B. **PIBID/Ciências Biológicas/CAC-UFG: Initial Impressions**. In: Proceedings of the 6th[a] Congress of Teaching, Research and Extension, 2009, Goiânia-GO.

PARAMETROS CURRICULARES NACIONAIS: **Introduction to the National Curriculum Parameters**. Secretariat of Fundamental Education. 2.ed. Rio de Janeiro: DP8, 2000.

PAVANI, J.; MOURA, Gutemberg. **Geographical, Urban and Architectural Panorama of Boa Vista**. Brasília: Coronário, 2006.

PIRES, Célia Maria Carolino; CONDEIXA, Maria Cecília; NÓBREGA, Maria José M. de; DIAS, Paulo Eduardo. **Symposium 20:** for a curricular proposal for the 2nd segment in the EJA. In: Congresso Brasileiro de Qualidade na Educação Formação de Professores - Simpósios, (org.) Marilda Almeida Marfan, vol.1, Brasília, 2002.

PORTAL DO GOVERNO DO ESTADO DE RORAIMA, **Education in the State of Roraima**. Available at: <http://www.portal.rr.gov.br/index.php?option=com content&task=view&id=492&itemid=1>. Accessed on: 13 Apr. 2011.

REVISTA, Jornal Folha de Boa Vista - **Boa Vista 118 years, in facts and photos**, published on 9 July 2008. Located at Rua Santos Dumont, Sn, São Francisco, Boa Vista, RR, 2009.

RIGAL, Luis. **The critical-democratic school:** a pending issue on the threshold of the 21st century. In: Education in the 21st century: the challenges of the immediate future, ed. Porto Alegre: Artes Médicas Sul, 2000.

RIGAL, Luis. **"La escuela popular y democrática: un modelo para armar".** In: Education in the 21st Century: The Challenges of the Immediate Future, ed.
Ernani Rosa. 2.ed. Porto Alegre: Artes Médicas Sul, 2000.

SECRETARIAT OF BASIC EDUCATION. Human Sciences and their Technologies.
Curriculum guidelines for secondary education. Brasília, v. 3, 2008.

UMBELINO, Ariovaldo. **Education and geography teaching in the Brazilian reality**. Paper published in the journal Desalambar, n° 6. AGB. Brasília: DF, 1987.

VESENTINE, José William. **Teaching geography in the 21st century.** São Paulo: Papirus, 2004.

APPENDICES

GOVERNMENT OF THE STATE OF RORAIMA

STATE UNIVERSITY OF RORAIMA - UERR

GEOGRAPHY DEGREE COURSE

GEOGRAPHY QUESTIONNAIRE

REGULAR SECONDARY SCHOOL STUDENTS

Date: __________________________

Sex ( ) Female ( ) MaleAge   **:** ________________________

Class: _____________________________________ **Shift:** ____________________________

School: ____________________________________ **City:** ________________________

Public ( )   **Private ( )**

( ) How would you define the subject of GEOGRAPHY today?

( ) 1 - important for your own knowledge.

( ) 2 - it's a very broad discipline of knowledge.

( ) 3 - is present in everything that happens in the world.

( ) 4 - it is a discipline that is constantly changing.

( ) 5 - it's just an easy and decorative subject, with no importance for academic and professional growth.

( ) All alternatives 1 to 4.

2)   How is the relationship between the GEOGRAPHY teacher and the students?

( ) great

( ) good

( ) regular

( ) bad

3)   Can you understand the content presented by your GEOGRAPHY teacher?

( ) yes   ( ) no

4)In   relation to the GEOGRAPHY teacher in the classroom, what are the NEGATIVE points?

( ) Hasn't mastered the geography content.

( ) Doesn't know how to explain the geography content.

( ) talks about other subjects in geography lessons.

( ) They only pass tests in geography lessons.

( ) is only late for geography lessons.

( ) doesn't teach the 60-minute geography lesson.

( ) only uses the geography textbook.

( ) Doesn't do didactic lessons, such as classroom dynamics.

( ) I don't have a degree in geography.

( ) All of the above.

5)   What are the POSITIVE points about the GEOGRAPHY teacher in the classroom?

( ) has a degree in geography.

( ) Can explain the geography content.

( ) uses tools other than the geography textbook.

( ) teaches didactic lessons, such as dynamic classroom lessons.

( ) does field lessons in geography.

( ) gives seminars to students in geography lessons.

( ) teaches the 60-minute geography lesson.

( ) is punctual in geography lessons.

( ) Has mastered the content of geography.

( ) All of the above.

6) What suggestions would you make to the GEOGRAPHY teacher for developing work in the subject?

( ) field lessons in geography.

( ) classroom dynamics in geography.

( ) The use of maps in geography.

( ) The use of technology in geography.

( ) Working on research in geography.

( ) Working together with practice and theory in the subject of geography.

( ) use tools other than the geography textbook.

( ) All of the above.

7) How important is GEOGRAPHY in your intellectual and professional development?

( ) Important for my intellectual formation.

( ) Important for my professional training.

( ) Important for my intellectual and professional development.

( ) No importance for my intellectual formation.

( ) No importance for my professional training.

( ) No importance for my intellectual and professional development.

( ) I don't need to study geography.

( ) The study of geography is essential.

UERR
UNIVERSIDADE ESTADUAL DE RORAIMA

GOVERNMENT OF THE STATE OF RORAIMA

STATE UNIVERSITY OF RORAIMA - UERR
GEOGRAPHY DEGREE COURSE

QUESTIONNAIRE ON THE INSTITUTIONAL
TEACHING INITIATION SCHOLARSHIP PROGRAMME
(PIBID)

<u>REGULAR SECONDARY SCHOOL STUDENTS</u>

Date: ___________________________

Sex ( ) Female ( ) MaleAge   : ___________________________

Class: ___________________________________ Shift: ___________________________

School: _________________________________ City: _____________________

Public ( )   Private ( )

1) Are you aware of the Institutional Teaching Initiation Scholarship Programme (PIBID) at your school?

( ) Yes   ( ) No

2) Do you know the importance of the Institutional Teaching Initiation Scholarship Programme (PIBID) in your school?

( ) Yes   ( ) No

3) Do you know why your school was selected to take part in the Institutional Teaching Initiation Scholarship Programme (PIBID)?

( ) Yes   ( ) No

4) Did you like the work done by the geography students through the Institutional Teaching Initiation Scholarship Programme (PIBID) at your school?

( ) Yes   ( ) No

5) In your opinion, was the relationship between the students and the academic fellows of the GEOGRAPHY course good?

( ) great

( ) good

( ) regular

( ) bad

6) In your opinion, has there been any progress in teaching and learning after the activities developed by the academic fellows in the subject of GEOGRAPHY through the Institutional Teaching Initiation Scholarship Programme (PIBID)?

( ) Yes   ( ) No

7) Could you identify the activities carried out by the PIBID GEOGRAPHY scholarship holders that were most important?

( ) Field class;

( ) classroom dynamics;

( ) research work;

( ) theory and practice;

( ) Classes with fun activities;

( ) Building models;

( ) Other.

8) Would you like the activities of the Institutional Teaching Initiation Scholarship Programme (PIBID) developed in the subject of GEOGRAPHY to remain at the school?

( ) Yes   ( ) No

9) Did you become interested in the subject of geography after the PIBID/GEOGRAFIA programme started at your school?
( ) Yes   ( ) No

10) In your opinion, what actions or activities should be developed by PIBID/GEOGRAPHY scholarship holders in order to improve the teaching-learning process?

____

____

____

____

ANNEXES